KX4Z Sound Card Interface Kit Construction Manual

Gordon L. Gibby KX4Z

Version 1.1   April 2017

ISBN-13: 978-1545480076

ISBN-10: 1545480079

Version 1.1 April 19, 2017

Cover Design
First Prototype sound card interface constructed & worked well. Small improvements have been made with the Version 1.1 printed circuit board described in this booklet..

PREFACE

This instruction manual is designed to accompany a printed circuit board created using ExpressPCB printed circuit board design tools. I developed this printed circuit board to:

- dramatically speed up construction of transformer-isolated sound card interface circuitry
- allow the use of laptop-built-in sound card, or inexpensive external USB-based sound cards, some of which function well with Raspberry Pi `linbpq` and `alsamixer`
- considerably reduce the cost of getting into digital amateur radio communications modes, compared with (otherwise excellent) commercial isolation interface systems

For the amateur radio club or group that wishes to turn this into a project, I'll gladly send them (by email) the .pcb file that allows them to order their own boards. While finished boards start at about $20, they drop to $10 with large quantities. (Contact me if you would like to join in an order.) The parts for this kit can be obtained under $10. A simple steel electrician's house wiring 2-gang junction box can be had for just over a buck, and a cover for just over 50 cents. Makes for a very inexpensive shielded box! With an experienced and patient mentor, this would be a great project for helping amateur radio operators with little previous construction experience, learn what fun it can be to actually *build your own equipment.*

Contact me at: docvacuumtubes@gmail.com

Gordon L. Gibby MD KX4Z
Newberry, Florida
March 28 2017

REF: https://www.expresspcb.com/

Gordon L. Gibby

DEDICATION

This short instruction manual is devoted to my long-suffering and wonderful wife, Nancy Gibby, who put up with the kitchen table being taken over many, many times during the development of this system....my own workbench already covered with other projects.

She's a great example of a Lady who serves her Lord.

And the verse below (written 2,500 years ago), was used by Pastor David Chauncey of Westside Baptist Church, Gainesville, FL to urge Christians to be *active* in serving their communities.

*Seek the welfare of the city where I have sent you into exile,
and pray to the Lord on its behalf;
for in its welfare you will find your welfare.*

Jeremiah 29:7
c. 627 BC

CONTENTS

Chapter	Title	Page
1	Construction, connection & adjustment	1
2	External Connections	23
3	Mounting / Enclosure Options	33
4	Adjustment	39
	Appendices -- Reference Information	41
	About the Author	45

Gordon L. Gibby

ACKNOWLEDGMENTS

The brave souls of the Alachua County ARES group who spent HOURS soldering together kit after kit, back before we had this printed circuit board, and who put up one after another BPQ node stations, deserve a huge amount of appreciation. And John Wiseman G8BPQ for writing wonderful "BPQ" software that has benefited so many. And then Andrei Kopanchuk, UZ7HO, of Ukraine, who has greatly served the ham radio community by writing wonderful software that does packet decoding and terminal functions, soundmodem and easyterm.

Read MORE About Amateur Radio Digital & Voice Emergency Communications in my other book:

**AVAILABLE ON AMAZON
AS PAPERBACK & KINDLE**

Figure -- The version 1.1 Printed Circuit Board that makes constructing this kit so much easier.

1 CONSTRUCTION

NOTICE: THIS KIT REQUIRES UNDERSTANDING OF SIMPLE CIRCUITRY AND BASIC SOLDERING AND CONSTRUCTION ABILITIES. IT MAY NOT BE APPROPRIATE FOR BEGINNERS WHO DO NOT HAVE ADEQUATE MENTORING. THE INSTRUCTIONS ARE SET AT A LEVEL APPROPRIATE FOR A BUILDER WHO HAS BUILT KITS BEFORE. IF YOU HAVE NEVER BUILT A KIT BEFORE, SEEK A COMPETENT MENTOR BEFORE BEGINNING. PLEASE READ THE ENTIRE INSTRUCTIONS BEFORE BEGINNING. THIS KIT AND THE INSTRUCTIONS AND ANY OTHER COMPONENTS ARE PROVIDED WITHOUT ANY WARRANTEE OR GUARANTEE OF PERFORMANCE AND THE BUILDER AGREES TO HOLD THE DEVELOPER OF THIS KIT HARMLESS FOR ANY AND ALL DAMAGES.

INTRODUCTION
In order to be able to transmit information digitally using a voice transceivers (whether SSB on HF bands, or FM on VHF/UHF bands) some circuity/software is needed to change alphanumeric characters into tones that will go into a microphone jack, and to key the transmitter. Vice versa, to unscrambled the mishmash of audio signals and turn them into readable data, software/hardware is needed also.

Hardware-only solutions (which often included a processor running obligatory software, but hidden inside) have long been sold and named Terminal Node Controller, or PACTOR modem, or similar. Recently, the power of the personal computer and capabilities of modern sound cards have taken on much of this work, so that a much simpler hardware system, providing adjustment of signal level, isolation of transceiver from computer, and push-to-talk (PTT) activation, can be used along with relevant software. Commercial solutions abound for this, with the Tigertronics Signalink being very popular, along with the Rigblaster, a recent MFJ product, and some older systems that relied on a signal from a 9-pin serial port.

Representative (AND INCOMPLETE) Collections of Software / Hardware Digial Solutions

Frequency Bands	Digital Mode / Goal	Software Required (Example)	Hardware Required
HF	Winlink / WINMOR for email	RMS EXPRESS[1]	sound card and interface capable of electronic control of PTT
HF	PSK31, Olivia, etc. used for QSO's	FLDIGI[2], Ham Radio Deluxe or any of multiple software packages.	sound card and interface; PTT can be done manually if electronic control not available
VHF/ UHF	Packet / email	RMS EXPRESS +	sound card, software and

---

1 Download from www.winlink.org
2 Download the version for your operating system from: https://sourceforge.net/projects/fldigi/files/fldigi/

		soundmodem.exe[3]	interface capable of electronic control of PTT
VHF/UHF	"Classic Packet" where you use a keyboard to contact other hams or stations, very much like FLDIGI does.	Easyterm.exe[4] terminal software plus Soundmodem.exe (Note: error-corrected; includes file transfer capabilities.)	sound card, software and interface capable of electronic control of PTT

The cost can be reduced somewhat by building one's own interface, and either using a laptop's built in sound card, or an inexpensive USB-connected sound dongle. For the Raspberry Pi user (including those building a `linbpq`-based node) a "classless" (no driver required) USB audio adapter such as the Adafruit 1475 (typically $5[5]), Sabrent USB sound card[6] or equivalent is preferred. These sound card are "classless" and do not need any additional driver

---

3 Download soundmodemXX.exe and easytermXX.exe from http://uz7.ho.ua/packetradio.htm (XX = version)

4 Download easyterm.exe from http://uz7.ho.ua/packetradio.htm; uses AGWPE interface, works well with soundmodem.exe from same developer.

5 Available from: https://www.adafruit.com/products/1475 as well as from: https://chicagodist.com/search?q=1475

6 Available from https://www.amazon.com/gp/product/B00IRVQ0F8/ref=oh_a ui_detailpage_o05_s00?ie=UTF8&psc=1

software. The <u>classless adapters</u> will function properly with Raspberry Pi (linux) Raspbian `alsamixer` to allow easy gain adjustment of the sound card system; generally you want to have fairly high signal levels in and out of the sound card (but not into distortion!) to reduce the damaging effects of hum pickup etc. Then inside the sound card adapter, gain adjustments are made with a hardware potentiometer. If you're using `alsamixer`, make your adjustments permanent afterwards with

```
sudo alsactl store
```

Figure 1-1 (below) shows a schematic of a simple transistor-based sound card interface that provide for gain adjustment in both directions, transformer-based ground isolation, and fast automated PTT control derived from the audio on one channel (VOX-type circuit). The basic idea of this circuit is certainly not original with me; countless previous similar designs have been presented. It simply consists of transformer isolation of signals in both directions between the sound card and the radio transceiver, and a simple audio amplifier driving a diode detector to create a VOX-type signal that is then used to switch a relay, as well as PushToTalk indicator LED to facilitate initial setup.

In my experience, the relay is key in avoiding RFI-induced tranmitter-locked-on hangs. Opto-isolation would be expected to be similarly successful. Direct transistor output control in my experience was more problematic, which is why I switched to relay output for PTT control.

Sound Card Interface Kit

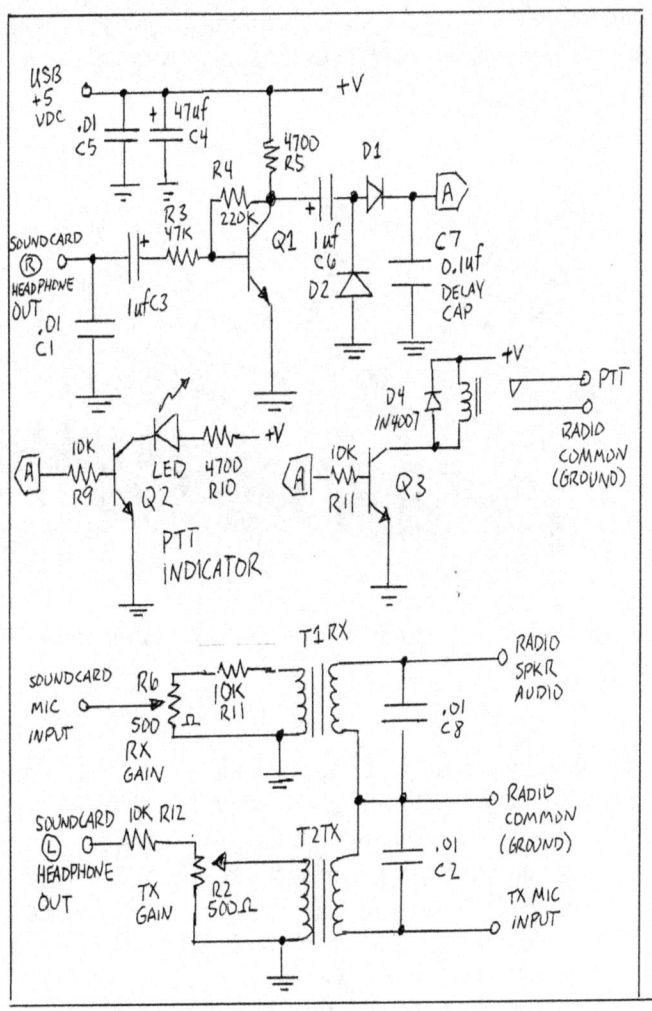

FIGURE 1-1. *Schematic for sound card interface. The values of R11 and R12 (4.7k - 10K) are not critical but may be changed to better match your radio's signal levels; 10K will give you a little finer control of signal levels. The*

printed circuit board includes an additional "backward" diode between USB+5 and Ground to protect against accidentally applied reverse power supply voltage.

Figure 1-2   V*ersion 1.1 of printed circuit board for interface circuitry.*

Improvements made to the circuit in the V1.1 board (which are not yet reflected in the schematic above):
- Added "backwards diode" D3 between +5 and ground to protect against reverse supply polarity
- Placed 22 ohm 1/4 watt "optional" series resistor in +5 line to act as a current limiter/fuse in the event of a short circuit within the circuitry. It is recommended to use this series resistor.
- Optional 0.01 uf capacitors across the sound card mic input and left channel headphone output to further reduce RFI. These aren't recommended unless you experience significant RFI. I haven't ever needed them.

The author & many friends have built a dozen or more of this circuit and they are in continuous usage inside several packet node stations, as well as in use for WINLINK client email applications, as well as casual radio QSO's. However, construction on standard perfboard takes an experienced builder roughly 2 hours, and novices may take *quite a few* hours.

To make construction much faster and easier, a printed circuit board design was created which resulted in a 2-layer PCB (3.8" x 2.5") with silk screen lettering.

CONNECTIONS:

This circuit connects sound card signals, arranged generally along the left side of the board, to an amateur radio transceivers' mic/speaker/ptt, which are connected generally at the upper middle and right hand side of the board as follows:

SOUND CARD CONNECTIONS

Connection	Location
sound card ground	bottom left corner of board, two larger pads, marked USB GND
USB +5 volts	Use the "optional" 22 ohm series resistor in the +5VDC line, and connect the USB +5V to one of the "+5VDC IN" pads above & to the left of the 22 ohm resistor.
Sound card	2 larger pads upper left

mic	corner labeled "USB MIC"
Left channel headphone output (used to send signal for transmission to the transceiver)	2 larger pads just below the mic input, labeled "USB L-CH"
Right channel headphone output (used to operate the PTT via a "voice-activated VOX" type circuit	2 larger pads left lower portion of the board, labeled "USB R-CH"

RADIO CONNECTIONS

Connection	Location
Radio ground (*isolated from USB ground*)	Center of boxed-in area, center-top of board, labeled "gnd". Multiple pads provided
Radio MIC input	Lower portion of boxed-in radio connection area, labeled "mic". 2 pads provided
Radio SPEAKER OUT	Upper portion of boxed-in radio connection area, labeled "spkr". 2 pads provided
Radio PTT connection (relay output, limit current to 50 mA). During intended transmission, this connect is shorted to Radio ground.	Right hand edge of the board, labeled "RADIO PTT". 2 pads provided.

NOTE: if your radio has separate PTT and MIC grounds, this circuit does not provide separated grounds for those purposes, so connect both of them to "Radio Ground."

Extra pads were provided for connections simply for redundancy.

COMPONENTS

The components in the table that follows will be required. All resistors can be 1/4 watt. (The board is drilled for 1/4 watt resistors.) All electrolytic capacitors can be 10V or greater (e.g., 15, 25 or 35 will work). Transistors are specified as 2N3904 but many general purpose small NPN transistors would work. Know the emitter, base, collector pinout of the transistor you are using! (They are NOT all the same!)

Figure 1-3. *2N3904 pinout with flat side up, curved side down.   TO-92 package.*

The relay is a somewhat delicate but fast reed relay. Most radios have a tiny push-to-talk current, well within the reed relay's capabilities. If you're using an older tube-type transceiver, check to see that the PTT current doesn't exceed 50 mA. The specified relay has 4 pins on 0.2" centers, while the pc board has 6 pads on 0.1" centers. **Thus not all the pads will be used!** The pads are wired so that the relay may be installed starting either from the bottom pad or the top pad (there will be one left over at the opposite end) and it works either way. This was due to the limited number of standard "components" offered in the design package and my lack of knowledge to create a customized relay pad structure. The relay is symmetrical, it can be installed with either end up or down. The center two pins are the coil, and the end pins are the relay contacts. The Ver. 1.1 screenprint puts circles on suggested insertion points; the relay is symmetrical and can be installed with either end upwards.

Figure 1-4. *Reed Relay. Device is symmetrical, center pins are the coil, end pins are the contacts.*

Either of two physical sizes of trimmer potentiometer may be used. The tiniest size requires a thumbnail or small plus head screwdriver; the larger size can be operated with the fingers. Alternatively, wires can be soldered and go to a panel mounted potentiometer, if for example you'll be using this device with multiple radios. For most installations connected long term to a single radio, simple set and forget will work well.

The transformers have one winding to the left, and the other to the right. The circuit board center hole is not used on each side (not connected). I'm unable to tell primary from secondary on general purpose 600 ohm 1:1 transformers, so mount either way as long as one side is LEFT and the other side if RIGHT, not up and down.

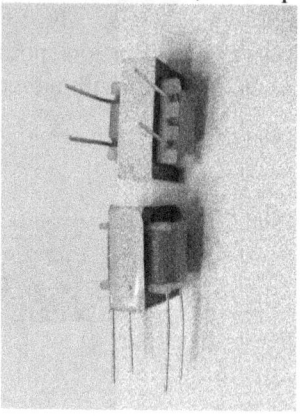

Figure 1-5 *Audio 1:1 Transformers. Yours may have center taps (not used in this circuit) or may be as these, with only 2 connections on each side.*

The printed circuit board has positions for optional 0.01 uf capacitors across sound card mic and left headphone channels.   I haven't needed these, but they may benefit some.

u = "micro"

NOTE: *Most of these components are literally only PENNIES. The cost of shipping is one of the larger costs, and frequently if you buy 10 of an item you get a price break. As an amateur radio operator who is going to the trouble to order parts, you might want to buy some "extras"* *and keep them in your "spare parts" drawer.*

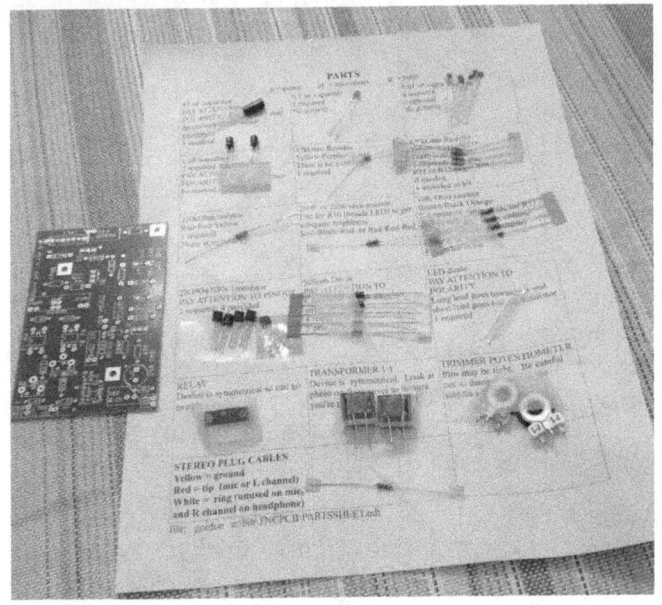

Figure 1-6 *Complete set of printed circuit board parts for one interface. There are extras of some of the parts included here.*

Component	Qty req.	Digikey Part No.
0.01 uf ceramic capacitor (used to filter out RF)	4-6	490-11884-ND https://www.digikey.com/product-detail/en/murata-electronics-north-america/RCER71H103K0K1H03B/490-11884-ND/4277785
0.1 uf capacitor to act as short	1	445-2637-ND https://www.digikey.com/product-

delay capacitor		detail/en/tdk-corporation/FK26X7R2E104K/445-2637-ND/970587
1 uf electrolytic capacitor	2	493-10230-1-ND https://www.digikey.com/product-detail/en/nichicon/UMF1V010MDD1TP/493-10230-1-ND/4312489
47uf electrolytic capacitor to filter out any hum on +5 line	1	P5539-ND https://www.digikey.com/product-detail/en/panasonic-electronic-components/ECA-1EHG470/P5539-ND/245138
22 ohm 1/4W resistor, optional, used as current limit/fuse in the +5 line	1	CF14JT22R0CT-ND https://www.digikey.com/product-detail/en/stackpole-electronics-inc/CF14JT22R0/CF14JT22R0CT-ND/1830311
47K 1/4W resistor	1	CF14JT47K0CT-ND https://www.digikey.com/product-detail/en/stackpole-electronics-inc/CF14JT47K0/CF14JT47K0CT-ND/1830391
4700 1/4W resistor	2-4	CF14JT4K70CT-ND https://www.digikey.com/product-detail/en/stackpole-electronics-inc/CF14JT4K70/CF14JT4K70CT-ND/1830366
10K 1/4 W	4	CF14JT10K0CT-ND

Sound Card Interface Kit

resistor		https://www.digikey.com/product-detail/en/stackpole-electronics-inc/CF14JT10K0/CF14JT10K0CT-ND/1830374
220K 1/4W resistor used to bias Q1 into quasi linear range	1	CF14JT220KCT-ND https://www.digikey.com/product-detail/en/stackpole-electronics-inc/CF14JT220K/CF14JT220KCT-ND/1830407
2N3904 transistor epoxy package TO-92 case Pay attention to the mfgr. drawing of EBC leads.	3	2N3904FS-ND https://www.digikey.com/product-detail/en/fairchild-on-semiconductor/2N3904BU/2N3904FS-ND/1413
Diode, 1N4007 or almost any diode	4	1N4007-TPMSCT-ND https://www.digikey.com/product-detail/en/micro-commercial-co/1N4007-TP/1N4007-TPMSCT-ND/773694
500 ohm trimmer	2	Choice of miniature or thumb-adjustable size Miniature: 3306K-501-ND https://www.digikey.com/product-detail/en/bourns-inc/3306K-1-501/3306K-501-ND/84791

			Thumb adjustable: 201XR501B-ND https://www.digikey.com/product-detail/en/cts-electrocomponents/201XR501B/201XR501B-ND/98331
LED (push to talk indicator)	1		C503B-RCN-CW0Z0AA1-ND https://www.digikey.com/product-detail/en/cree-inc/C503B-RCN-CW0Z0AA1/C503B-RCN-CW0Z0AA1-ND/1922930
Reed Relay	1		306-1062-ND https://www.digikey.com/product-detail/en/coto-technology/9007-05-00/306-1062-ND/301696
600 ohm 1:1 audio transformers	2		These can be obtained much more inexpensively over ebay. The impedance can be anything near 600 ohms. Here is an example of TEN transformers for less than $4, shipped from Europe: http://www.ebay.com/itm/10X-Audio-Transformers-600-600-Ohm-Europe-1-1-EI14-Isolation-Transformer-TSUS/112271494516?_trksid=p2045573.c100506.m3226&_trkparms=aid%3D555014%26algo%3DPL.DEFAULT%26ao

		%3D1%26asc%3D41376%26meid%3Dd49cdfe8fb154623a304652fcb7f689c%26pid%3D100506%26rk%3D1%26rkt%3D1%26 Digikey has an expensive model at $5.60 each: MT4135-ND https://www.digikey.com/product-detail/en/tamura/TTC-105-1/MT4135-ND/285702
Snap on ferrite core for cables	2	240-2599-ND https://www.digikey.com/product-detail/en/laird-signal-integrity-products/28A1507-0A2/240-2599-ND/2242762

Figure 1-7. *Built and installed Version 1.1 board, with wiring including 2 (black) audio cables, USB cable, and shielded ethernet radio cable. Note screws and insulating wood paneling underneath printed circuit board.*

SOLDERING & CONSTRUCTION

I'll never forget one of my friends trying to solder tiny 1/4 watt resistors and heat-sensitive 2N3904 transistors using the blunt tip of a huge 100-watt soldering gun! *There are ways to make this project difficult, and that is one of them!*

Use a low wattage soldering iron, such as 20-25 watts, with a fairly sharp and well-tinned (with solder) tip. While lead-free solder is encouraged today, 60/40 lead/tin rosin-core solder (still available) melts at a very low temperature, flows well and is very easy for beginners to work with. Try to avoid breathing the fumes!

The usual way to build a printed circuit board is to install a few components (3-5), bend their leads just a bit so that the component hugs the surface of the board, and then quickly solder each lead on the BOTTOM of the board with a well-applied iron and a quick touch of solder. As soon as the solder melts and flow, remove both iron and solder to avoid over-temperature. Solid state devices --- particularly transistors, can be damaged by too high a temperature for too long, so solder fast -- only a very few seconds are required. Then clip off the excess leads (avoiding hitting an eye with the projectile lead). Move on to the next components. (Do not solder on the top of the board.)

The following quasi-step-by-step instructions are provided as a help:

1. Mark on the board the two 0.01 uf capacitors at the upper left side between the sound card connection pads and the potentiometers -- don't initially install these unless you end up with intractable RFI (not likely).
2. Resistors may be the easiest components to mount. So start with a few of these -- at the lower left end of the

board, there are 47K, 4700, 220K, and the "optional" 22 ohm resistor near the 4700, which you can install to gain practice soldering on this board.

3. Electrolytic capacitors require that you place their - and + sides correctly. There are two 1-uf and one 47-uf electrolytics to mount -- inspect the parts to see if their negative (-) or positive (+) leads are marked and install and solder properly. If you find the negative side marked, then of course the positive is the opposite lead.

4. Diodes are the next most important item to get right, but it is easy -- there is a "bar" on the symbol on the circuit board, and one of the actual diode itself. Mount so that they match. Avoid overheating them.

5. The transistors are thermally the most fragile item -- and you must get their orientation correct to have the emitter, base, and collector leads in the right holes. Observe the rounded and flat sides of the TO-92 case

Insert the transistor so the leads are in the proper holes (Figure 1-7 may help) and solder with a bit of skill so that it is done quickly.

6. The transformers must not accidentally be installed with their cores 90 degrees off -- observe them in Figure 1-7 for proper mounting.

7. The relay is symmetrical and its mounting has been

discussed above.

8. For the TX and RX gain potentiometers, holes were provided for both "small" and "large" sized trimmer pots -- use whichever fits best.

9. There is no polarity required on the 0.01 and 0.1 uf capacitors, nor is there on any of the resistors.

10. The PTT (push-to-talk) LED does have a polarity -- the cathode side is marked on the board with a bar, and the actual device has a shorter lead and a "flat" on the case on the cathode side.

11. Install the remainder of the components. R11(RX) and R12 (TX) can initially be 10K resistors; if RX (receiver) and TX (transmitter) signals aren't strong enough, change the relevant one to 4700 ohms (doubling the signal available).

12. Refer to Chapter 2 for external connections to both the sound card side, and the radio transceiver side.

13. Near the drilled optional mounting hole on the right hand side of the board are two pads that can be connected with a spare length of component lead or wire, in order to connect the USB circuit ground (not the transceiver ground) to the mounting screws, and hence to a metallic case, if desired. This may improve RFI protection.

INITIAL SAFETY TESTING

Before applying power, very carefully look through the entire circuitry to be sure that you have put the right components in the right places, have polarized components (such as diodes, transistors, electrolytic capacitors, LEDs) inserted properly. This circuit has a "fuse" made of a 22 ohm "optional" 1/4 watt resistor to try and avoid damage to the USB bus of a laptop. When power is first applied, do it for for only a second or two and watch for signs of untoward effects. It is normal for the LED to briefly flash when power is applied. Add a few seconds each successive connection, until the circuitry has proven itself safe.

The collector of Q1's voltage (versus USB-ground) should be measured using a voltmeter. It should neither be saturated (0.2 V) nor cut-off (5 volts) -- it should best be somewhere in the middle (1-4 volts). If that voltage isn't right, something isn't correct with the biasing of Q1. Check:
 a) right resistors in right spots?
 b) solder connections all good?
 c) +5 supply is really +5?
 d) transistor wasn't fried during soldering?
....then consider getting help from a mentor.

Once the circuit is deemed safe, connect receiver audio (see Chapter 2 for full connection information) and verify that audio proceeds to the computer through the mic input (usually allowing signals to be seen on a waterfall, and characters displayed on a monitor). Adjust the RX GAIN if nothing shows up.

Next work on testing the PTT circuit, applying audio to the R CHAN in-- the PTT LED should illuminate. BE CERTAIN THAT YOUR LAPTOP SPEAKER OUTPUT TO THE RIGHT CHANNEL IS AT 100% to properly activate the PTT LED & relay.

2 EXTERNAL CONNECTIONS

SOUND CARD SIDE

The circuit requires connections to the headphone output and microphone input of a sound card, which can be the internal one of a modern laptop, or a USB-connected sound card dongle. Additionally, the circuitry requires a +5VDC source, which can be obtained from a USB port. (You could alternatively use any 5V source, or even up to a 9V battery.) For the beginner, this is the easiest method and will be discussed first. The Ebay kit includes 2 stereo 1/8"/3.5 mm audio cables/plugs for this purpose and a USB cable for obtaining the +5 VDC.

STEREO AUDIO CABLE /USB PLUG METHOD
First obtain two 3.5mm (1/8") stereo plugs with cables, preferably using normal wire, not the tiny nylon-embedded wire filaments with painted insulation used with some audio cables. Determine which wire represents ground, tip and ring of the jacks. The color codes in the table below apply to some batches of mass-produced cables -- but check with an ohmmeter to confirm the proper wires of your kit.

Microphone Plug	Typical color code
Tip = microphone input	red
Ring = not used	white
Sleeve = ground	yellow

Headphone Plug	Typical color code
Tip = left channel	red
Ring = right channel	white
Sleeve = ground	yellow

Wire the appropriate wires to the correct pads on the left hand side of the printed circuit board. You will probably want to mark the plugs so that you know which one to plug into mic and which to headphone output. Red tape or paint on the MIC plug is common.

USB Plug +5VDC Connection

This is utilized merely to gain the +5V needed to operate the circuitry. Using any standard (type A) USB cable, open up the cable to gain access to the wires. These connectors are fairly standardized and the color code is usually:

USB CABLE COLOR CODE
Black =	Ground
Red =	+5V
White	Data
Green	Data

While avoiding shorting any of the colored wires to any other of the wires, use a voltmeter to verify the presence of +5V on the red wire while plugged into a USB receptacle.

With the proper wiring verified, connect the +5V to the left side of the 22 ohm "optional" resistor used for fusing purposes in this project. Connect the ground wire to the USB ground pads at the lower left portion of the printed circuit board.

The audio and USB cables can pick up RFI (radio frequency interference), and conduct it into the sound card, resulting in stuck-on transmitter, or locked-up computer or sound card. Putting 2-3 coils of 2" diameter in these cables may reduce the pickup, as well as adding a ferrite bead around the cables.

Skip to the RADIO SIDE in this chapter.

DIRECT WIRING METHOD

As an alternative, direct (soldered) connections can be made to the internal microphone input, headphone outputs (both L & R channels), ground and +5V of an inexpensive Adafruit $5 USB audio adapter (sound card). My preferred sound dongle for this connection method is the Adafruit 1475. Typically when this alternative is employed, the Adafruit audio adapter is deposited within the same shielded box with the audio interface printed circuit board, and a USB extension cable is used to connect the short USB cable from the Adafruit audio adapter to the computer. A shielded USB extension cable is preferable. Significant small-circuit soldering skills are needed for this approach. The advantage is the lack of need for several cables, and the placement of the Adafruit sound adapter inside the shielded enclosure further reduces RFI.

When the white case is carefully opened up with a knife tip or tiny screwdriver, the necessary wiring connection points are as shown in Figure 2-1. Use appropriately tiny flexible stranded wire. The most difficult connection is the +5VDC because the USB wire tends to become unsoldered while you are trying to tap in with your added wire. A bit of finesse can make this succeed.

The connections to the mic and headphone soldering jacks are much easier. The metallic connections on the jacks readily accept soldered wires. The grounds on both headphone and mic connector appear to be DC connected together, so either be used as the ground connection.

Once you have made the necessary connections, you may wish to secure the board and some of your wiring with a generous drop of 5-minute epoxy. A 1/4" hole drilled into the removed cover will allow the wires to be routed out of the plastic enclosure, which can then be reassembled (with glue or other fastener as required)

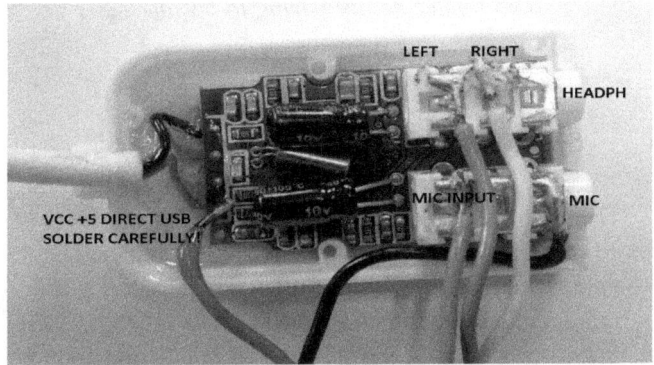

Figure 2-1. Close-up of direct soldered connections to Adafruit 1475 sound adapter, after gently prying case open. The +5V red wire connection requires special finesse to avoid the delicate red USB wire on the underside coming loose. The right-most top connection of both headphone output (top) and mic input (bottom) are connected together in the 1475 sound adapter.

Figure 2-2. Completed V 1.0 (prototype) circuit board

wired (by direct soldered connections) to an Adafruit 1475 sound adapter. (The connections to the sound adapter can also be made by stereo 3.5 mm (1/8") cables/plugs[7], and USB +5 and ground obtained using a cut-off USB cable (RED generally +5V and BLACK = ground, but verify with a voltmeter before wiring to be safe.)

RADIO SIDE

The simplest method to connect to your transceiver is to solder four stranded wires from a multi-wire shielded cable to the mic / ptt / receiver audio / ground pads at the center and right hand side of the board (see Figure 2-2), and then terminate the cable with the appropriate mic connector for your radio, in some cases needing also a 1/8" plug to obtain receiver audio output from an "external speaker" jack. This method is reliable and works fine if you are going to use the sound card interface with only one radio.

The Ebay kit includes a couple feet of ethernet twisted-pair cable (usually shielded) for this purpose. Typically it is suggested to use the following colors for the radio signals:

Signal	Wire Color
Mic	White-Orange
Ground	Orange (and bare shield wire)
PTT	White-Green
Rx Audio	White-Blue

---

[7] Stereo cable terminated with 3.5mm stereo plugs at both ends, can be found for less than $1 such as: https://www.monoprice.com/product?p_id=644 Cut in half to make both mic and headphone portions from one cable.

If you may utilize more than one radio, and need to be able to make connections to radios with varying connectors, you might prefer to provide an intermediate disconnection point. The popular Signalink accomplishes this with a female RJ45 jack on its rear panel. You can emulate this with a surface mount type RJ45 jack, wired to the radio connections of the printed circuit board.[8] Alternatively, you can terminate the radio cable from the PCB with a male RJ45 and convert to a jack using a double-female RJ45 jack, which are less than 50c online[9] Then different cables, each with an RJ45 plug on one end, and the appropriate connector(s) on the other end for each radio, can be prepared for each different radio.

The question arises: what pins in the intermediate RJ45 shall be used for which signals? In the popular Signalink product, this is changeable for different radios to match commercially available radio cables. **However this introduces an unnecessary unknown (the pinout of the RJ45 jack) into equipment for emergency communications that would be best standardized.**

Our emcomm group in Alachua County standardized on one pinout, so that volunteers would always know the signal pinout on the sound interface card RJ45 plug, and as much as possible, we also use the same pinout on all Signalinks in use; it is the one recommended for popular Baofeng UV5RA

---

8   Surface mount ethernet jack:
    https://www.amazon.com/Monoprice-Surface-Mount-Single-107090/dp/B005E2Y9RY/ref=sr_1_1?ie=UTF8&qid=1491853472&sr=8-1&keywords=surface+mount+ethernet+jack
9   https://www.monoprice.com/product?c_id=105&cp_id=10519&cs_id=1051902&p_id=7280&seq=1&format=2

radios, and also for mini-6-pin DIN radio connections.

ALACHUA COUNTY FLORIDA STANDARDIZED RJ45 PINOUT

PIN	SIGNAL	Wire Color
1	microphone	white/orange
2	ground	orange
3	push to talk	white/green
4	unused	blue
5	receiver audio	white/blue

This table assumes the numbering shown in Figure 2-3, where the gold pins are held upwards and away from the reader.

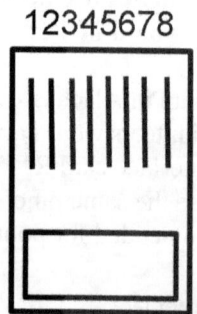

Figure 2-3 *Top View (pins visible) numbering of the RJ45 plug pins.*

This happens to be the standard pinout of commercially available cables intended to connect the popular Baofeng

UV5RA (which uses a special molded double-plug connector) to a Signalink. This allows one to easily plug in a Baofeng UV5RA low power transceiver to test a system.

You can also purchase surface mountable 8-pin RJ45 jacks which could be mounted on your enclosure. My local Home Depot carries those.

RFI REDUCTION

(A bit of trial and error, usually, and VERY important.)

To reduce RFI, put 3 or 4 loops (2-3" dia.) in the audio cable from the radio to the interface circuit board, and also clip on a ferrite "bead" if possible.
You may also put 3 or 4 loops (2-3" dia.) in audio cables to the USB sound card dongle if it is external to your metal enclosure.
If you used the direct-solder-to-sound-card technique and thus use a USB-extension cable from an internally mounted sound card dongle to your laptop, also put loops in that cable, and a ferrite "bead".
Laptop touchpads may frequently experience RFI when you touch them while transmitting. In that case, use a wireless mouse.

Try to use well-balanced external antennas on your transceiver; handi-talkie mounted rubber-duckie antennas are the worst, as they are inherently unbalanced and often create large unbalanced RF currents on audio wiring connected to the handi-talkie.

3 MOUNTING/ ENCLOSURE OPTIONS

To protect the soldered connections on the bottom of the circuit board, cut off unnecessary lead length after soldering, and affix a piece of cardboard either with a bit of epoxy glue or (better) some double-stick mounting tape such as used to hold posters to walls. This will avoid short circuits from metallic surfaces touching the bottom of the circuit board. Instead of cardboard, you might want to use a thin piece of wood (such as 1/4" balsa wood or wood paneling) if you are going to mount the circuit board with screws which might compress cardboard.

It is advantageous to mount this circuit inside a metal box of some sort to provide RF shielding. Computer and transistor circuitry can be very sensitive to radio frequency energy picked up by wires and leads. Such a shielding box can be constructed in several ways, from the extremely cheap to the very expensive

1. **Extremely Cheap:** Enclose in a cardboard box, bring out the required leads, cover the box with simple aluminum foil like a Christmas present, and

tape. The aluminum doesn't even necessarily have to be connected to the ground of the circuit, though that might help.

2. **Attractive** Enclose in an appropriately sized metal tin such as for fine teas, or wallets or other small commercial items. This can make a very attractive exterior. Be careful of sharp edges if you drill holes for the leads to come out. Rubber grommets can be purchased (Internet or hardware store) or a "non-metallic" electrician box cable clamp can be used--- these are usually very inexpensive.

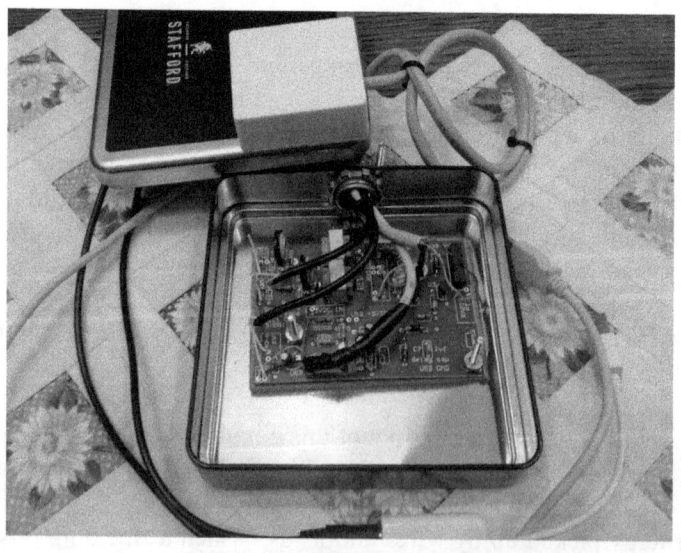

Figure 3-1. *Mounted using screws to a left-over metal box. A layer of 1/4" spare wood paneling was used as an insulator to protect the solder connections from shorting out to the metal box. The attractive metal box was the packaging of a nice man's wallet received as a Christmas present. Electrician's 1/2" non-metallic cable clamp is used to secure and protect the cables passing through hole*

drilled in metal box using 5/8" wood borer. 6-32 screws can be used.

3. **Utilitarian** (My preferred) use a 4" electrical junction box with blank steel lid, and 3/8" ("non-metallic") clamp connector to protect the leads from sharp edges. Such a box already includes knockouts, which can allow you easy access to the trimmers by removing the appropriate knock-out. The board JUST fits into a 4" junction box. (A tiny bit of sanding on the edge of the board might make it fit better.) These are available inexpensively from any home improvement store in the $1 range. Use double-stick mounting foam tape to secure the board into the junction box. Sand or file any rough edges on the flat cover that matches the box.

Figure 3-2. *4" Electrical box, blank steel lid, and two*

different sizes of "non-metalllic" cable clamps. The cable clamps are made to protect non metallic cables from being cut by sharp edges, and also clamp them to protect internal circuit from tension. The steel box has "knock-outs" to provide holes of various sizes, pre-made. These boxes are made in various depths, and are quite inexpensive compared to hobbyist aluminum boxes. Note this one has rounded corners. Add felt stick-ons on the bottom, and sand or file any sharp edges.

Figure 3-3. Printed circuit board mounted (actually wedged) into a 4" electrical box. One knockout has been removed to allow access to the potentiometers, The USB dongle is also inserted, and another knockout has been

removed and a "3/8 non-metallic clamp" used to secure the wires' exit point. There is a standard solid metal plate that fits on top as a cover.

4. **Most Expensive:** Purchase an aluminum hobby box (e.g., from Digi-key or Mouser). The production board includes 3 drilled holes with solder flats. If you use metallic standoffs, and wish to connect the box electrically to the USB ground, there is a jumper position that can be used to do that at the lower right hand side of the board.

It may help to connect the USB ground to the metal enclosure, but it isn't always necessary.

Gordon Gibby KX4Z

4 ADJUSTMENT

This assumes that you are familiar with the software you will be using (e.g., EasyTerm, WINLINK, linbpq, BPQ32, etc.). Select the proper choice to pick the sound card that is connected to the interface. With the squelch on your receiver wide open, adjust the receiver volume and the RX GAIN trimmer for the best results on character detection while listening to an active packet channel if possible. Some radios provide a direct connection to the demodulator that will have a constant 100 mV (or similar) signal level, independent of the transceiver's front panel speaker volume level adjustment; this is ideal.

The goal in adjusting the transmitted level is to get just below the maximum correct deviation of your transmitter -- which for FM corresponds to an audio perceived signal in an FM receiver that is slightly softer than maximum loudness. Cause your transmitter to transmit repetitively (e.g., calling a non-existent station) and adjust for just below the maximum deviation while listening on a 2nd transceiver.

For HF SSB, adjust so that you are near the top of the linear range of output power but NOT so as to "flat top" or cause significant Automatic Level Control (ALC) to be developed by your transmitter. (This advice may need modification for some models, however.) You can observe your transmitted signal level in any of several ways, possibly including an display of your transmitted power, an analog forward SWR meter, or a power meter in the transmission line.

MORSE CODE

At modest speeds you can even use the PTT to key a CW transmitter if desired. Be certain that the current carried is not more than about 50 mA and that the open circuit voltage isn't above approximately 15 volts. The reed relay was not meant to switch significant power. FLDIGI will happily send and receive CW for you, and by using the relay PTT output to key a CW transmitter, there is question that this is legal in the appropriate segments of the HF band even for Technician licensees. Alternatively, you can use the audio output (L channel) to send modulated tone CW, which is very similar to A0 if there is little hum or distortion on your output sine wave.

NOTE: Why does this circuit require both left and right audio channel input, and uses the RIGHT channel to run the PTT? Why not simply detect the presence of audio on the L channel? The reason is that some modes (particularly some HF modes) vary the signal amplitude significantly and might not keep the PTT activated -- thus FLDIGI provides the option of a helpful continuous tone on the R channel to guarantee steady PTT. Packet doesn't have this problem! For packet, you could actually tie both L and R Channel sound inputs together and drive with only one channel of your laptop, potentially allowing you to run two different speeds/modes with two circuits.

APPENDICES

REFERENCE INFORMATION

MICROPHONE JACK PINOUT AS VIEWED FROM THE EXTERIOR OF THE TRANSCEIVER

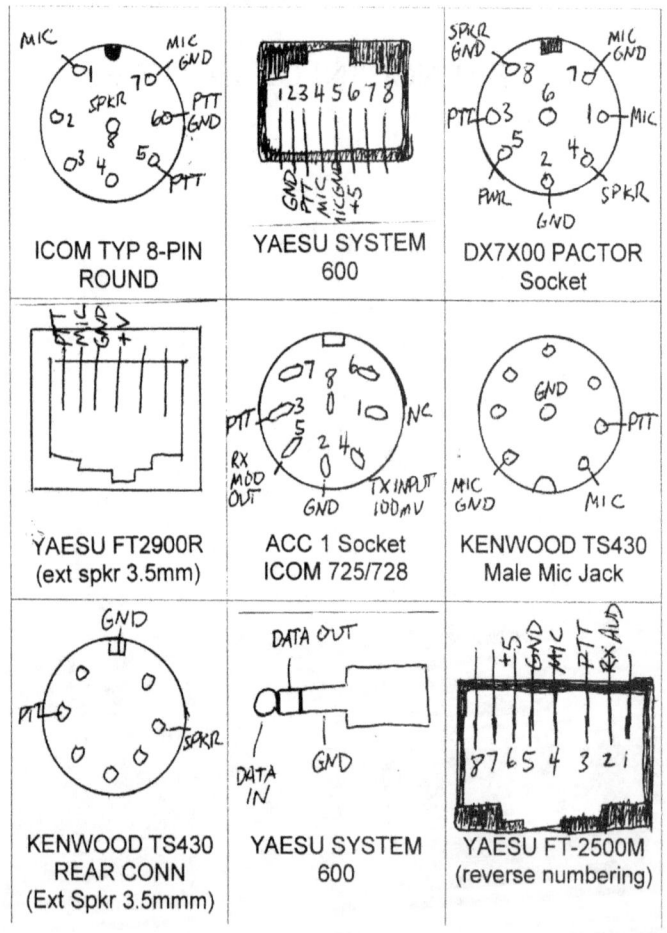

Sound Card Interface Kit

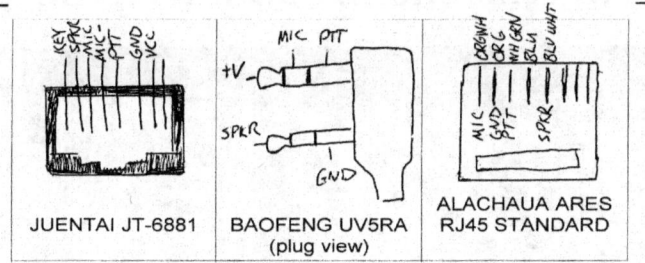

PRINTED CIRCUIT BOARD LAYOUTS

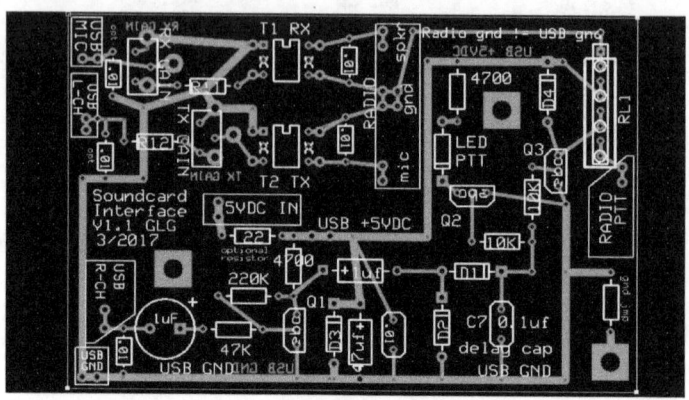

(For the Kindle viewers) Color photo of bottom copper (green) and top lettering (yellow)

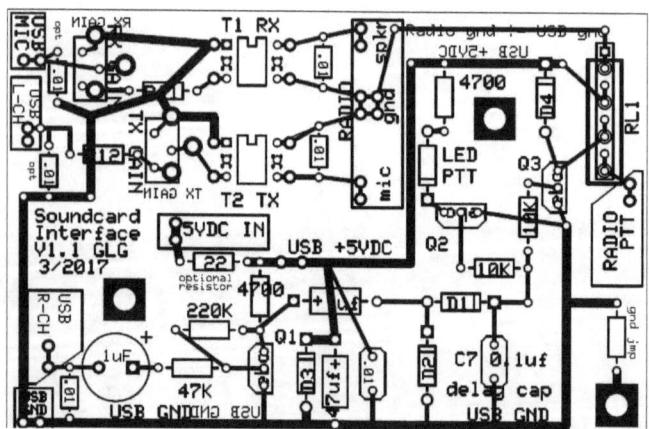

For the print readers -- black and white printout of the bottom copper and top screenprinting. Lettering that is "mirror image" is in copper on the bottom.

ABOUT THE AUTHOR

Gordon L. Gibby is a practicing physician with previous education in Electrical Engineering. As kids grew up and moved off, more time became available for amateur radio and other hobbies to be resumed. As a high school student, he learned construction from venerable Heathkit vacuum tube transceivers (one of which is still in routine usage -- even for Winlink!) as well as high power vacuum tube amplifier design.

His current amateur radio interest is Emergency Communications, particularly the high speed data communication possible with modern digital software.

Gordon Gibby KX4Z

Read MORE About Amateur Radio Digital & Voice Emergency Communications in my other book:

AVAILABLE ON AMAZON & KINDLE

www.ingramcontent.com/pod-product-compliance
Lightning Source LLC
Chambersburg PA
CBHW061221180526
45170CB00003B/1104